CAREERS IN

MICROBIOLOGY

THE WORLD IS FULL OF LIVING THINGS so small that most of us will never see them. Yet these invisible organisms have significant effects on human life – some good, some bad. Some microorganisms can cause infections that pose serious health risks, and left unchecked can grow into a plague. Other kinds of microorganisms are beneficial, and can fight infection or "eat" toxic waste. Microbiologists are the scientists who study these microscopic life forms, and their interactions with people, plants, animals, and the environment.

This is a fascinating career, but it is not for "germophobes." That is because microbiologists deal with life forms most of us would rather not see: viruses, bacteria, parasites, and fungi. There are millions of different species! Every day, microbiologists are traveling to all corners of the earth and beyond to discover strange, new forms of life that no one even knew existed. Their job is to identify microbes, determine whether they can help or hurt us, and develop ways to either utilize them for our benefit or combat them for our protection.

Microbiologists work in almost every industry from agriculture, oil, and pollution control, to biotechnology, pharmaceuticals, and healthcare. They work for government agencies, academic institutions, corporations, technology firms, and industrial and environmental organizations. Most work in sterile research labs or hospitals, but some, like environmental microbiologists spend their time outdoors, slopping through muck and mud to get samples and conduct field experiments.

The education required to work as a microbiologist depends upon the position you are aiming for. Some microbiology laboratory technician jobs, particularly in medical laboratories, are open to individuals with a high school diploma or an associate degree. But generally, a bachelor's degree is the minimum requirement to enter the field, and top research positions and jobs at colleges and universities often require a graduate degree, either a master's or doctoral.

For those who have obtained a solid education, the job outlook is terrific. The need for qualified microbiologists is growing at an accelerated pace, both for basic research and practical applications. No one knows how many microbes are yet to be discovered. More importantly, microbiologists are still learning how the known microbes function and how they can be utilized for the benefit of society. As a result, the field of microbiology has unlimited employment potential.

The field of microbiology is loaded with attractive

features. The pay is good, with the potential to earn over $100,000 a year. There are so many more reasons to consider this career. Microbiologists are paid to discover things in a world most people will never see, and unlock some of nature's best-held secrets. They get to travel the world, enjoy prestige among their colleagues, and have a newly discovered species named after them. It is a field that is wide open, and there is no better time to take a look at the amazing work you could be doing.

START YOUR EXPLORATION NOW

HIGH SCHOOL STUDENTS ASPIRING TO become microbiologists should plan to take as many science and math classes as possible. Striving for good grades in courses such as calculus, physics, computer science, chemistry, and biology is the best way to prepare for a science major in college. To stand out on college applications, enroll in as many AP courses as possible. English and speech classes are important because microbiologists are often called upon to write articles and papers, and give presentations and speeches. Foreign languages are also helpful because microbiologists routinely exchange information with other scientists around the world.

Talk to your science teachers about your interest. They can help you learn more about the field in a variety of ways. First, they can help you find scientists in your area that you can talk to about what microbiologists actually do. They can set up job shadowing opportunities, arrange tours of laboratories, lead field trips to science museums, and recommend books you should read.

Find out if your school or community has a science club, science enrichment programs, or other activities you could get involved in. Participate in science fairs and competitions, like Science Olympiad.

Do volunteer work over the summer in a research laboratory to get as much practical and realistic experience in the field as possible. Hands-on experience will provide a good basis for making college and career decisions.

Join local and national scientific professional societies, like the American Society of Microbiology (ASM). Participate in their online forums and live events.

HISTORY OF THE PROFESSION

MICROBIOLOGY HAS HAD A LONG, rich history. Although the existence of microorganisms was merely a hypothesis for many centuries before their actual discovery, scientists believed long ago that there was unseen microbiological life all around us. In fact, as early as the 6th century BC, the Jain scriptures described sub-microscopic creatures living in large clusters and having a very short life. These *nigodas* were said to pervade each and every part of the universe, including tissues of plants and animals.

The ancient Romans also believed in creatures so small they could not be seen with the eye. Roman scholar, Marcus Terentius Varro, even warned against locating a homestead in the vicinity of swamps because they were ideal breeding grounds for minute creatures that could cause serious diseases. In the medieval Islamic world, scientists spoke of microorganisms that carried contagious disease such as tuberculosis. Girolamo Fracastoro claimed in 1546 that epidemic diseases were caused by seed-like entities that could not be seen.

All these early claims about the existence of microbes were based on nothing more than speculation. Until the invention of the microscope in the 17th century, there was simply no way to observe, prove, or accurately describe them.

Historians are unsure who made the first observations of microbes, but an English scientist named Robert Hooke was the first to write of key observations. He reported observing strands of fungi among the specimens of cells he viewed. It was a humble beginning for modern microbiology. However, a Dutch merchant named Anton van Leeuwenhoek, who died in 1723, expanded Hooke's work greatly during the late 1600s. Van Leeuwenhoek made careful observations of microscopic organisms he called *animalcules*, which were actually bacteria. Although he was an amateur microscopist, he is credited for providing the first accurate descriptions of protozoa, fungi, and bacteria.

Despite Leeuwenhoek's findings, microbiology did not really make any great progress for quite some time. Microscopes were rare and there was little interest in studying something that could not easily be seen. In 1872, Robert Koch demonstrated that certain organisms cause disease. His experiments, known as Koch's Postulates, formed the foundation for modern microbiology. Yet, the scientific community at large did not accept his work until Louis Pasteur demonstrated his anthrax vaccine in 1882. Still, many scientists refused to accept the existence of microorganisms. It wasn't until the 19th century when other sciences produced higher quality lab equipment, including better microscope lenses, that interest in the field began to grow.

Pasteur showed that heat could be used to destroy microbial cells, and Koch developed the first methods for the growth of pure microbial cultures. These discoveries enabled others to produce a number of important practical advances and brought about the revolution in molecular biology. Pasteur is most famous for his series of experiments designed to disprove the theory of spontaneous generation. He is also well known for designing methods for food preservation (pasteurization) and vaccines against several diseases such as anthrax, fowl cholera, and rabies. His contemporary, Koch, is best known for his contributions to the germ theory of disease, which proved that specific diseases were caused by specific pathogenic microorganisms. His work led to the discovery of several unique bacteria, including *mycobacterium tuberculosis,* the causative agent of tuberculosis.

In the 20th century, the field of microbiology expanded to include virology. Work with viruses could not be effectively performed until instruments were developed to help scientists see these disease agents. In the 1940s, the electron microscope was developed and perfected. In that decade, cultivation methods for viruses were also introduced, and the knowledge of viruses developed rapidly. With the development of vaccines in the 1950s and 1960s, such viral diseases as polio, measles, mumps, and rubella came under control.

In the second half of the 20th century, more than 2,000 species of animal, plant, and bacterial viruses were discovered. They ranged from the common rhinovirus that causes colds, to deadly versions like SARS and HIV. Despite the achievements of scientists over the past one hundred years, viruses continue to pose new threats and challenges.

The mid-20th century is considered the "Golden Age of Microbiology" because so many agents of different infectious diseases were identified. Many of the causes of microbial disease were discovered, which led to the ability to halt epidemics by interrupting the spread of microorganisms. However, it was not until after World War II that antibiotics were introduced in medical care. The incidence of pneumonia, tuberculosis, meningitis, syphilis, and many other diseases declined with the use of antibiotics. At the same time, scientists found better methods for cultivating viruses – a necessary step in the development of vaccines. In the 1950s and 1960s, many serious viral diseases such as polio, measles, mumps, and rubella came under control.

In recent years, microbiology has branched into multiple disciplines. For example, one of the major areas of applied microbiology is biotechnology. In this discipline, microorganisms are used as living factories to produce pharmaceuticals that otherwise could not be manufactured. These substances include the human hormone insulin, the antiviral substance interferon, numerous blood-clotting factors and clot-dissolving enzymes, and a number of vaccines. Bacteria can be re-engineered for many purposes, such as increasing plant resistance to insects and frost, editing DNA, and producing sustainable energy. The possibilities seem infinite, and biotechnology will certainly represent a major application of microbiology throughout the 21st century.

Microbiological research will continue at a rapid pace for many years to come. Scientists will refine techniques and tools, and learn how to manipulate more and more of the microbial world for our benefit. Clinical and public health scientists will need to be vigilant as new diseases emerge and old ones (such as tuberculosis, typhoid, and diphtheria) re-emerge in vaccine resistant strains. The need to preserve the environment will lead to the investigation of novel ways of using microbes to clean up pollution. Biotechnologists will be motivated to find ways to mass-produce materials and medicines cheaply and cleanly. As the understanding of microbial genetics increases, and as more genomes are sequenced and mapped, the ability to manipulate, combat, and utilize microbes will increase.

WHERE YOU WILL WORK

MICROBIOLOGY IS A DIVERSE FIELD that offers employment opportunities in dozens of areas in both the private and public sectors. Employers include the federal government, state and county health departments, and academic institutions. Microbiologists are also employed by pharmaceutical companies, biotechnology firms, food and beverage industries, and manufacturing companies. Academic microbiologists often teach classes, in addition to conducting their own research.

Organizations that research and develop products in physical, engineering, and life sciences employ the largest share of microbiologists. In fact, one out of four works in this industry. Nearly the same number of microbiologists work for pharmaceutical companies that research, develop, and manufacture medicines. The federal government is another major employer. Nearly 15 percent of all microbiologists work in federal agencies such as the Food and Drug Administration, US Department of Agriculture, National Institute of Health, NASA, Environmental Protection Agency, and all branches of the military. Large numbers of microbiologists can also be found working for colleges, universities, and professional schools as well as state government agencies.

There are also numerous employers that hire smaller numbers of microbiologists, including:

Food and beverage processors

Hospitals

Biotechnology firms

Oil industry

Diagnostic laboratories

Bioremediation companies

Agricultural industry

Independent research institutes

Aerospace research and manufacturing companies

Most microbiologists work indoors in laboratories and offices, where they conduct research and write up the results of their experiments using computers.

Not all microbiologists spend their days inside sterile surroundings with their pipettes and microscopes. Environmental and medical microbiologists often work outdoors. Those working in agricultural and veterinary fields also spend time outdoors, collecting samples in the field. Because microbes are found throughout the world, this fieldwork can be carried out in challenging and remote locations. Microbiologists working in the field have to be able to survive with difficult living conditions – mud, snow, slime, and water.

Microbiology is not the type of work that can be done independently. Regardless of employer or industry in which they are employed, microbiologists work on teams with technicians and other professionals. It is routine to collaborate with other scientists from many different disciplines. Supervising more inexperienced scientists and students working in their laboratory is a common duty of fully trained professionals.

Contract work, rather than full-time employment, is becoming more common in this occupation. This trend is creating even more opportunities to select your type of work and location. More microbiologists are choosing to travel far from home to gain new perspectives and to experience different cultures and work styles.

THE WORK YOU WILL DO

MICROBIOLOGISTS ARE SCIENTISTS who study the structure, form, behavior, growth, and distribution of microbes, as well as their relationships and interactions with other organisms. They study the different ways microbes affect animal, plant, and human life in the environment. Professionals in this field design research projects and conduct experiments to learn more about the microscopic world.

The work of microbiologists may seem to be strictly academic, but it is not. In fact, there are far more microbiologists working on practical applications than there are working in pure research. Some microbiologists look for ways to use microbes to improve drugs, food, or the environment, while others work to control the spread of disease. For example, a doctor may ask a microbiologist to determine which microbe is present in a patient's bloodstream. A government agency overseeing water quality might ask for the identification of a pathogen in a body of water. A drug company may need help in isolating microorganisms from soil to create new antibiotics.

Most microbiologists work as part of a team. An increasing number of scientific research projects involve multiple disciplines, and it is common for microbiologists to work closely with chemists, biochemists, geneticists, pathologists, physicians, environmental scientists, engineers, veterinarians, and geologists. For example, microbiologists researching new drugs may work with medical scientists and biochemists to develop new medicines, such as antibiotics and vaccines.

Microbiologists work in almost every industry and have many different responsibilities. Because there are so many different species of microbes out there, and they do many very different things, no one microbiologist can cover the entire spectrum. That is why people who become microbiologists usually focus on a particular microbe or research area. The following is a partial list of roles that often overlap. Depending on the specific situation, a microbiologist may perform more than one function or role.

Bacteriologist

Bacteriologists study bacteria, particularly in relation to medicine and agriculture. They investigate various properties of bacteria such as growth, metabolism, diversity, and evolution. They also research both the positive and negative effects bacteria have on plants, animals, and humans.

Clinical Microbiologist

Clinical microbiologists study the various clinical applications of microbes for the improvement of health. Specifically, they research microorganisms that can cause, cure, or be used to treat diseases in humans. There are four kinds of microorganisms that cause infectious disease: bacteria, fungi, parasites, and viruses. It is the job of the clinical microbiologist to determine which microbe might be the cause of infections in humans and animals, and what antimicrobials may be effective for treatment. These professionals play a vital role in the detection of new infectious agents.

Immunologist

Immunologists study how the human body defends itself again disease. They study how immune systems react to and fight invading microorganisms.

Mycologist

Mycologists study the properties of fungi, such as yeast and mold, as well as the ways fungi can be used to benefit society. For example, fungi are used to make medicines such as penicillin or in the production of food and beverages such as beer, wine, cheese, and edible mushrooms. Mycologists also study the dangerous aspects of fungi, such as poisoning or infection.

Virologist

Virologists study the structure, development, and other properties of viruses, as well as how they infect and affect host organisms such as the human body. Viruses are submicroscopic, parasitic particles of genetic material contained in a protein coat. Virologists generally focus their attention on the diseases viruses cause, techniques to isolate viruses and culture them, and how they can be used in research and therapy.

One of the basic tasks of the virologist is to classify viruses. Thousands of viruses have been identified, but there are countless more, each of which can infect all types of life forms.

Biotechnologist

Biotechnologists manipulate genes in order to modify microorganisms. Their work sometimes creates controversial new products for human use, such as genetically modified organisms (GMOs) found in processed foods. The work can be beneficial, such as crops that are drought resistant, or drugs that fight disease that would otherwise go unchecked.

Microbiologist

Microbiologists examine microbes using a variety of tools, such as epifluorescence and confocal microscopes that allow them to see what the human eye cannot. They may use other equipment, including gas chromatographs, mass spectrometry, fermentors, autoclaves, electrophoresis gels, fluorescent cell sorters and various incubators. Some laboratories use toxic chemicals or dyes when working with microbes.

Across the broad range of specialties available in microbiology, many positions have similar job activities. In general, here are typical duties of microbiologists.

Collect samples from the environment, including people, plants, animals, and field locations

Grow microbes from samples, using standard laboratory techniques to isolate specific microbes

Plan and conduct complex research projects, such as developing new drugs to combat infectious diseases

Identify and classify microorganisms found in specimens collected from humans, water, food, and other sources

Monitor the effect of microorganisms on plants, animals, and other microorganisms and on the environment

Keep up with findings from other research groups by reading research reports and attending conferences

Prepare technical reports, research papers, and recommendations based on research findings

Present research findings to scientists, business executives, engineers, government agencies, other colleagues, and the public

THE MANY FACES OF MICROBIOLOGY

MICROBIOLOGISTS ARE WORKING everywhere. This huge and ever-expanding profession has found its way into every private and public sector you can think of, simply because microbes and their actions pervade all aspects of our increasingly complex society. Here are just a few of the areas where the need for microbiologists has exploded in recent years.

Medical Applications

Medical microbiologists help scientists and physicians in the diagnosis, prevention, and treatment of infections in animals and humans. In general, the work involves investigating how organisms cause disease, and their role in the disease process. Some medical microbiologists specialize in studying the occurrence of disease in a specific population, and how to control epidemics. This specialty is known as *epidemiology*.

Some medical microbiologists do not conduct research, but instead work in healthcare, providing help with diagnostics and quality control in hospitals. Others create, develop, and check the quality and safety of vitamins, antiseptics, vaccines, antibiotics, and other medicines for companies in the pharmaceutical industry.

Applied Science

Applied microbiology is a term that refers to the development of products for a company. Also known as *industrial microbiology*, professionals in this specialty can be found working for a wide range of companies, from big biochemical businesses to smaller firms that develop biopharmaceuticals or specialty products. Positions in industry include management, quality control, research, and product development. Although the most common companies are in big pharma and biotechnology, there are numerous other industries where microbiology plays a significant role. These include cosmetics and toiletries, breweries and other beverage producers, mining, oil production and refining, dairies, and food processors of all kinds.

Genetics

Geneticists study the process by which organisms inherit and transmit genetic information. The language of life is written with four letters, A, C, G, and T, representing the four nucleotides – adenine, cytosine, guanine, and thymine. These molecules linked together form deoxyribonucleic acids (DNA). Together the letters create a code of life. Geneticists are hard at work, cracking that code through the Human Genome Project, conducted by scientists around the world. They seek to explain the fundamentals and effects of heredity by studying the variations and commonalities between microbes and how they function.

Some geneticists are experimenting with genetic modification, while others are trying to understand how microbes survive in the most extreme environments. Some geneticists remain in academia doing their own research, but many more work in medical, agricultural or veterinary research, healthcare, forensics, brewing, pharmaceuticals, agrochemicals, and manufacturing, and in countless research institutes.

The Environment

Environmental scientists investigate the effects of microbial communities on the environment. Their studies provide information necessary for helping humanity cope with the consequences of modern life. The environment in this case generally means the naturally occurring soil, water, and air, as well as the plants and animals that inhabit Earth. Environmental microbiology also includes the study of microorganisms that exist in artificial environments, such as bioreactors.

Molecular biology has revolutionized the study of microorganisms in the environment and improved the understanding of the composition and physiology of microbial communities. Microbial life is amazingly diverse and microorganisms literally cover the planet. It is estimated that less than one percent of the microbial species on Earth have been identified so far, although environmental scientists have already discovered numerous microorganisms that can be used to combat pollution. In fact, microorganisms are not only cost effective, but also incredibly potent agents for remediation of domestic, agricultural, and industrial wastes, as well as hard-to treat subsurface pollution in soils, sediments, and marine environments.

Law

Legal experts with training in microbiology are in demand. The biologic revolution has resulted in countless discoveries that require industrial development and translation into products. Knowledge about microbial species, microbial products, and the application of microbes to industry is essential for the drafting of patents and technology transfer agreements. Microbiologists with knowledge of the law can be found in legal firms that specialize in patent law, technology transfer, and protection of intellectual property. These firms exist in all settings – academic, institutional, and governmental.

PROFESSIONALS TELL THEIR OWN STORIES

I Am a Bacteriologist

"I work in a university medical research lab. My job is to look at bacteria that cause infections in the stomach, intestines, and wounds. The work I do can make an impact on millions of people if I can figure out how to develop a vaccine or make a disease-causing microbe less harmful. I never touch a single patient, but what I do is essential to the advancement of medicine. Without research scientists like me, doctors wouldn't be able to cure people or prevent them from getting sick from harmful microbes.

I suppose I was always destined to be a scientist. A typical kid, I was fascinated by creepy crawlies and was constantly bringing home jars of bugs that I watched for hours at a time. The stranger the bug, the more excited I got. Today, I do the same thing, but with an electron microscope that magnifies invisible bugs (bacteria) several thousand times their actual size. It's intriguing to know that there is a world that none of us can see with the naked eye, yet it is all around us.

My childhood curiosity for what makes life work was quietly laying the foundation for my future as a microbiologist. My education towards that end, however, did not proceed smoothly. I took all the requisite math and science courses, but math was boring and I couldn't understand why I needed to learn chemistry when I was only interested in biology. It wasn't until my first college course in microbiology that I got it. Biochemistry and microbiology are intertwined and can't be separated. You must have knowledge of all the disciplines in the life sciences if you're going to be a successful researcher. Once I understood that, everything fell into place. I fell in love with microbiology and all its many facets.

Research work is exciting. I feel like a pioneer, charting new paths and writing new history. What research scientists like me do now will generate material for the textbooks of the future and help countless people live longer, healthier lives.

This has been a very rewarding career for me. I never look at the clock, hoping it is five o'clock so I can go home. I love my work and look forward to each new day. It isn't always easy. It requires a lot of patience and discipline, and the work is never-ending. Every answer to a problem opens up three or four more questions that need to be explored. Sometimes the questions are very difficult and it takes months or even years to answer them. The most successful researchers are persistent and determined.

I would suggest to anyone interested in the life sciences to consider a career in microbiology. There is so much work to do and the field is so open. As much as we have learned over the centuries, there is still much more that we don't know about microorganisms and their potential to benefit the lives of people everywhere."

I Work in a Hospital Medical Lab

"I always imagined I would be working in a research or clinical lab setting. In high school, I prepared by taking as many science and math courses as I could, including calculus, biology, chemistry, and physics. I didn't know anything about microbiology until my third year of college when I was suddenly immersed in it through classes in immunology, molecular biology, and other microbiology-related courses. It was fascinating and during the final year, I fell in love with it, when I did my clinical rotation that allowed me to work in a real lab.

A typical day varies, but I spend most of my time reading petri dishes and smears. I feel like a detective because I'm always trying to determine the type and amount of bacteria in a culture sample. It's not a glamorous job, but it is very rewarding. I am proud that my work is vital to the prevention, detection, diagnosis, and treatment of diseases that show up in this hospital. My role in the recovery of patients is never recognized because I work behind the scenes, but that is okay with me.

This is a good career for someone who has an aptitude for math and science, and is curious about other life forms – what they do and how they affect us, good or bad. It is not for the squeamish though. I work with various types of body fluids, and I am surrounded by numerous biological and chemical hazards. I have to pay close attention to safety guidelines, which can be a challenge when there are deadline pressures – as they always are.

My advice to someone interested in this field is to try it out before making any college plans. There are jobs you can get in a laboratory that don't even require a bachelor's degree. For positions like lab assistant and lab aide, you usually only need a high school diploma. Working in a lab environment will help you determine whether the profession is right for you.

Once you've made that decision to go for it, start

joining professional organizations. You should also get into the habit of reading professional magazines and scientific journals related to the field. That is the key way laboratory professionals keep up with what's going on, not just where you are, but all over the world. As microbiology knowledge advances, keeping up with current technology, protocols, and guidelines is critical to advancing your career. Belonging to professional organizations and volunteering on committees can help you network and gain valuable experience."

PERSONAL QUALIFICATIONS

WHAT DOES IT TAKE TO BE A MICROBIOLOGIST? In general, you must be smart and possess the practical skills necessary to do scientific research. There are also some very specific traits needed to succeed in the career.

Intellectual curiosity is key. Every successful microbiologist has an inquiring mind and a wide interest in natural phenomena, particularly at the microscopic level. These professionals are highly analytical. They use their critical thinking skills every day while searching for answers to complex questions.

Being smart is not enough. It is also important to be methodical and logical. Microbiologists spend most of their time conducting scientific experiments and analyses. The work must always be done with utmost accuracy and precision, but sound reasoning and good judgment are also needed to draw the correct conclusions from experimental results.

Microbiologists are very observant and vigilant. Experiments must be constantly monitored with an eye for the tiniest details. The best scientists are able to quickly and accurately identify microbes, and understand the significance of what is being observed. Drawing correct conclusions from what the microscope reveals can lead to significant scientific discoveries that could change people's lives.

While a sharp mind is certainly a prerequisite for working in this field, good hands are equally important. Manual dexterity is needed for mounting and staining specimens and transferring microorganisms from one culture medium to another without contaminating samples. The smallest slip-up could destroy months of important work. Developing a high level of skill with precision instruments is an ongoing challenge in the lab. An experienced microbiologist might be expert in the use of sophisticated microscopes, but this is a fast-moving field where new cutting-edge lab technologies are constantly being developed, refined, and upgraded.

Although microbiologists need to be able to work independently with minimal supervision, they do not usually work alone. Typically, they work on small research teams and therefore must be able to work well with others. Strong interpersonal skills are needed to motivate and direct other team members toward a common goal. Microbiologists must be able to interact with a variety of people while not getting caught up in competitive "brain wars."

Good communications skills are vital in this profession. Speaking skills are needed because microbiologists frequently give presentations about their work. They must be able to present concise and clear reports to their peers, and also be able to explain their research to people in the wider public arena who may be unfamiliar with the jargon. Microbiologists need strong writing skills, too. They are constantly writing memos, reports, and research papers that explain their findings. They also need to keep a complete, meticulous record of their work such as conditions, procedures, and results.

Microbiology work is not without its hazards. After all, microbiologists are dedicated to solving serious problems caused by bacteriological or viral threats. In the lab, safety concerns are paramount – in fact, they are subject to the most stringent safety regulations and practices for dealing with hazardous substances. Microbiologists must master the laboratory protocols and follow guidelines exactly to ensure a sterile work environment and prevent the accidental release of potentially disease-causing organisms or the contamination of sensitive experiments.

Perseverance and patience are virtues shared by every successful microbiologist. Scientific research involves substantial trial and error, and microbiologists must not become discouraged in their work. It can take months, often years, to complete some experiments.

ATTRACTIVE ASPECTS

BEING A MICROBIOLOGIST CAN BE very rewarding. You get to make discoveries for a living, in an amazing world that most people cannot and will not ever see. Whether you are working on a problem with human health or finding a new way to create sustainable power, the job can be stimulating and fun. It is exhilarating to break through a challenge and unlock some   of nature's secrets. Microbiology research has an impact on everything living on the planet – humans, animals, and plants.

Microbiologists can enjoy considerable prestige. When a project is successful, the research is published in well-known journals. Published scientists are often invited to speak at institutions all over the world. They write books and appear on TV.

If you find a new species, you can name it after yourself. This happens more often than you might think. It is estimated that the number of different bacteria in the world is five million trillion trillion – a five with 30 zeroes after it! Only a tiny fraction of existing species have been discovered. There is much work to be done.

There are opportunities to travel. Microbiologists often visit exotic and remote locations in the pursuit of their research. The world is your work place, as microbes have colonized basically all habitats on earth. Microbiologists sample more than microbes while traveling around. They taste exotic foods, experience new cultures, and collaborate with colleagues from other countries. They also travel throughout this country and abroad to attend important seminars, conferences, and workshops.

The field of microbiology is wide open. Little is known about microorganisms, yet they have the potential to revolutionize our way of life and the world we live in. It is a science that is growing exponentially, and the demand for young scientists is exploding. From soil microbial ecology to virology to molecular biology, there is virtually an unlimited number of research projects to choose from.

UNATTRACTIVE ASPECTS

IF YOU ASK MICROBIOLOGISTS WHAT they do not like about their career choice, they will be hard pressed to come up with a quick answer. Being a scientist can be rewarding in many ways, but no career is perfect and microbiology is no exception.

First negative to consider is that it takes many years of education and work experience to become a microbiologist. At the very minimum, you will need four years of college. You can land an entry-level job with a bachelor's degree, but the best jobs require two years of graduate work that result in a master's degree. However, even with a master's degree you will be very limited in the type of job you can get. Essentially, you will be working in a technical capacity, running the same tests over and over again without the potential thrill of discovery.

To achieve the most fulfilling career, you will need a doctoral degree (usually a PhD), which means four years of college followed by six years of graduate school. That is a decade of higher education!

If you like school, do well, and work very hard, you will be prepared to enter the working world. Finding the exact job you want may not be easy at first. While there are numerous opportunities, you may have to relocate to a city far from home or take a job in a specialty that does not interest you. The hard part is getting started. Once you get a job and show what you can do, doors start to open. After a few years on the job, you can be more selective in where you live and the work you do.

The job outlook for microbiologists is generally very good, but job security is better for some than others. If you decide to work in a lab doing research, for example, your future could be uncertain. Researchers have to write grant proposals constantly to compete for a source of funding to keep projects going, and only a fraction of grants receive ongoing financial support.

If you want to go into education, you will probably be middle-aged before you get your first tenure track position. This is a college professor who cannot be fired – usually Associate Professor level or higher.

Microbiologists make a comfortable living. However, if you go into this field for the money you will be disappointed. Scientists are at the lower end of the income scale when compared to other professionals with similar levels of education.

Although microbiologists work in clean, pleasant places, the work is not without risk. It is extremely important to be aware of possible chemical injury or exposure to infection when working with pathogens. Strict safety procedures and preventive inoculations help to protect medical microbiologists from the risk of disease.

EDUCATION AND TRAINING

BEGINNING MICROBIOLOGISTS usually complete at least a bachelor's degree program in microbiology or a related biological science. A bachelor's degree is sufficient to obtain some entry-level microbiologist jobs, but a doctoral degree is generally required to conduct independent research or teach.

College students hoping to become microbiologists do not necessarily have to major in microbiology. However, the degree should be in a subject closely related to biological science, such as biology, molecular biology, biochemistry, or chemistry.

A standard microbiology curriculum will include such courses as virology, pathogenic microbiology, immunology, microbial genetics, and bacterial physiology. Because it is important to have a broad understanding of the sciences and to be able to do complex data analyses, students also take classes in computer science, math, statistics, chemistry, biochemistry, and physics. The principles of microbiology are introduced in classroom lectures, and students are given the opportunity to gain firsthand laboratory experience in university facilities. Ambitious students often pursue research assistant positions to gain a better understanding of microbiology and improve their chances of getting into a graduate degree program.

Graduate School

A master's degree in science is required for many microbiology jobs, including those in applied research, inspection, and product development. A master's degree program is designed to provide specialized knowledge and experience in a particular field of study, while planning and conducting research experiments. Typical courses include biology of microorganisms, genetics of bacteria, advanced immunology, and bioinformatics. A master's degree in microbiology emphasizes research and usually requires completion of a thesis.

Earning a doctoral degree can qualify microbiologists for high-level research projects and top positions in private laboratories, university research departments, and government agencies. Many schools offer several different doctoral programs, each geared at a specialty within the field, or one of the many sub-disciplines. Programs are individualized, but the most popular studies focus on molecular microbiology, evolutionary microbiology, parasitology, microbial ecology, and neuro-immunology.

Upon graduation, these professionals may find employment with government agencies, university research departments and private laboratories. PhD programs usually require intensive classroom study, laboratory research, and completing a thesis or dissertation. Additionally, many students are responsible for planning and conducting research experiments on topics in their respective fields of study. Many prospective microbiologists engage in important research with professors and other students as part of a degree plan. It typically takes four to six years to complete a doctoral degree program in microbiology.

Postdoctoral Training

Many who have recently earned their doctorates begin their careers in a temporary postdoctoral research position, which typically lasts two to three years. In this job, they work with experienced scientists as they continue to learn about their specialties or develop a broader understanding of related areas of research. Some have the opportunity to publish research results, which can be very helpful when pursuing academic positions. A solid record of published research is essential to get a permanent faculty position in a college or university.

Laboratory Experience

It is important for prospective microbiologists to have laboratory experience before entering the workforce. Employers look for applicants with hands-on laboratory experience for microbiology positions at every level. Most college level microbiology programs include a mandatory laboratory requirement, but additional laboratory work is recommended.

Undergraduate students can hone their skills in internships with pharmaceutical companies or other employers while they complete their degrees. Both master's degree and PhD students normally conduct laboratory research for their dissertations. In addition, new PhD graduates usually spend at least two years after graduation in postdoctoral research positions to get additional experience.

Internships in this field are lengthy – new interns typically work with experienced microbiologists for several years. During this time, they help to organize experiments, apply for grants, set up materials, and record results. After gaining practical experience and producing meaningful research, the new microbiologist is often rewarded with the opportunity to conduct independent studies.

EARNINGS

NATIONWIDE, THE MEDIAN ANNUAL earnings for microbiologists are about $75,000. Those in the top 10 percent earn more than $125,000 a year, while those in the bottom 10 percent earn less than $50,000. Most of these professionals are earning between $65,000 and $95,000 annually. Incomes vary considerably depending on type of employer, geographic location, gender, and education.

Among employers, the federal government is the most generous. Microbiologists working for federal agencies earn about $100,000 on average. On the other end of the scale, companies that provide scientific research and development services employ the most microbiologists, yet the average pay is only about $70,000 a year. The second largest employer is the pharmaceutical and medicine manufacturing industry, which pays microbiologists even less – about $65,000 a year. Health clinics and animal food manufacturers do not employ as many microbiologists, but they offer above-average pay at about $80,000.

By state, microbiologists working in Maryland earn the highest average income, at about $100,000. Georgia ranks second for average pay, at $90,000. Louisiana is close behind, offering an average salary of $85,000.

Did you know that women outnumber men in undergraduate microbiology programs? There are more women holding PhDs in microbiology, too. But there does not seem to be strength in numbers. Although about 60 percent of undergraduate microbiologists are female, women earn median salaries of only $55,000 per year compared to their male counterparts who earn $65,000. The wage disparity eases with education and experience, but still, women at the top of their field earn slightly less than men with the same background.

Of all the variants, education has the most impact on earnings. According to a recent study by Georgetown University, the median salary for college graduate microbiologists is $60,000 per year. Not bad for a bachelor's degree, but obtaining a graduate degree raises income by almost 70 percent. That puts the median income for master's degree holders at just over $100,000 per year, according to this study, with the 75th percentile earning $140,000 yearly. It's no mystery why more than half of undergraduate microbiologists choose to pursue graduate degrees!

OPPORTUNITIES

THE FUTURE IS BRIGHT FOR ANYONE pursuing a career in microbiology. The number of jobs for microbiologists is expected to grow by almost 15 percent over the coming decade. Many new positions are being created in the pharmaceutical and environmental industries, in alternative energy, and in agriculture. Microbiologists who understand both microbiology and related fields such as biochemistry or medicine will have the best job opportunities.

The primary reason for such strong job growth is the need for new medicines, such as vaccines and antibiotics, to treat a growing and aging population. There is a growing demand for microbiologists trained in pharmaceutical and biotechnology research to develop biological drugs that are produced with the aid of microorganisms. Currently, these professionals are in a race against time to develop new antibiotics that can protect against "super bugs," aggressive bacteria that have become resistant to existing antibiotics.

Aside from improving our health, there is increasing interest in the development of clean, alternative energy sources, including biofuels and biomass – areas that require many more microbiologists. Agribusinesses will continue to employ microbiologists to develop genetically engineered crops that yield more produce with fewer pesticides and fertilizers, and require less water, for drought-stricken areas. There is increasing pressure to come up with better ways to clean and preserve the environment.

Microbiologists will find plenty of job openings in both private and public sectors. Industry remains a staple market for employment opportunities in applied microbiology. Industrial microbiologists may work in a range of companies where positions include management, quality control, research, and product development. Although many prospective applicants know of big pharma and biotechnology, there are numerous other opportunities ranging from cosmetics to breweries.

There are also numerous employment opportunities for microbiologists in government service. Microbiologists can be found working in many branches of governments. Concerns about bioterrorism have led to the hiring of microbiologists in law enforcement and military services. Government employment positions can be found at every level – local, state, and federal. Any agency that deals with public health, waste and wastewater management, food safety, or the environment is fertile ground for job seekers.

Even agencies that do not have an obvious connection to life sciences need microbiologists. NASA, for example, uses microbiologists to study factors such as infectious diseases, antibiotic resistance, and metabolic and genetic changes in the space environment. There are countless unknown microorganisms that already exist in space, plus space travelers bring bacteria, viruses, and other microscopic life forms with them from Earth. Once in space, these organisms react in unexpected ways to the environmental conditions that exist in reduced gravity or in the closed environment of a spacecraft. Future research will focus on the microbiological implications of long-term space travel and habitation.

GETTING STARTED

HOW DIFFICULT WILL IT BE TO LAND your first job with a Bachelor of Science degree in microbiology? That depends on how well you have done in school, your experience during your college years, and how willing you are to relocate. To increase your chances of finding the job you want, do well in your classes. Potential employers will look at your grade transcripts. Take advantage of internships, volunteer opportunities, and undergraduate research. Participation in these activities will set you apart. Above all, be flexible. Be willing to relocate and take a job that is not exactly what you want to do for the rest of your career. It is far easier to transition into a better position once you are employed than it is to find the perfect job the first time out.

Start your job search at your college placement office. There you will find a library of books and videos on job hunting, résumé writing, and interviewing. The placement office keeps a list of employers who will be visiting the campus throughout the year. You can request information on any of the companies that interest you so you can prepare for an interview. It is important to learn as much as you can about a company before you go to the interview. Sign up for as many interviews as possible. This is good practice. Each interview will give you more confidence and help you build your job-searching skills. If you need help, remember that your placement advisor is a professional who is probably the most qualified person on campus to lend assistance.

Extend your job search to Internet resources. You might find job openings advertised on general job boards, but microbiology is a relatively small universe. It is far better to visit professional association websites. For example, ASM Career Connections is ASM's online job board. It is dedicated to helping you find a job in the field of microbiology. You can post a résumé that is searchable by employers, search open job listings, contact employers directly from the site, and sign up to receive email alerts as new jobs are posted.

The Biotechnology Industry Organization also offers a career center on their website. Bio Jobs provides a wide range of information about biotech, including names, addresses, phone numbers, and other information for all kinds of biotech companies.

The value of networking cannot be stressed too greatly. In the scientific community, every professional you meet is a potential contact for a future job. This includes speakers at college seminars, college professors, internship supervisors, interviewers, fellow members of professional associations, and even other classmates. Keep notes for each person you meet – names, phone numbers, email addresses, what they do in their company, who they work for, how you met, etc. Create a LinkedIn account and add each one to your network. You can also search LinkedIn for new connections that you may never meet in person, but who could be helpful as you build your career.

ASSOCIATIONS

■ **American Society of Microbiology**
http://www.asm.org

■ **Society for Industrial Microbiology and Biotechnology**
www.simbhq.org

■ **IUMS International Union of Microbiological Societies**
http://www.iums.org

■ **American Institute of Biological Sciences**
http://www.aibs.org/home/index.html

■ **Federation of American Societies for Experimental Biology**
http://www.faseb.org

■ **Biotechnology Industry Organization (BIO)**
http://www.bio.org

PERIODICALS

■ **Microbe World**
http://www.microbeworld.org

■ **Microbiology**
http://mic.sgmjournals.org

■ **Journal of General Virology**
http://vir.sgmjournals.org

■ **Journal of Medical Microbiology**
http://jmm.sgmjournals.org

WEBSITES

■ **Microbiology Careers**
www.microbiologycareers.org

■ **Science Olympiad**
http://www.soinc.org

Made in the USA
Coppell, TX
22 November 2019

11703192R00020